**Happy Valley
Elementary School**
3291 Happy Valley Road
Victoria BC V9C 2W3

FAMILIES

Picture Credits:

Dwight Kuhn: pages 8-9, 15, 19
Robert & Linda Mitchell: pages 13, 14, 17, 19, 22
Nobert Wu: page 15
Roger De La Harpe/Animals Animals: page 24
M. Linley/Animals Animals: page 14
R. Schoen/Peter Arnold, Ltd.: page 25
Tom Brakefield/DRK: pages 23, 29
Stanley Breeden/DRK: pages 6, 11
Chamberlain/DRK: page 27
Michael Fogden/DRK: pages 8, 24, 29
Johnny Johnson/DRK: page 10
Stephen J. Krasemann/DRK: pages 10-11, 23, 26-27
Wayne Lankinen/DRK: page 21
Tom & Pat Leeson/DRK: pages 29
C.C. Lockwood/DRK: page 18
Allan Morgan/DRK: page 7
S. Nielsen/DRK: page 28
Doug Perrine/DRK: pages 26, 28
Anup & Manoj Shan/DRK: page 12
Peter Veit/DRK: page 17
Kennan Ward/DRK: page 7
J. Wengle/DRK: pages 20-21
Dick Dickinson/International Stock: page 6
Des & Jen Bartlett/National Geographic Image Collection: page 9
Jerry L. Ferrara/Photo Researchers: page 23
Jeff Lepore/Photo Researchers: page 7
Tom McHugh/Photo Researchers: page 25
Will Troyer/Visuals Unlimited: page 18-19
William J. Weber/Visuals Unlimited: page 13, 26
Michael E. Francis/The Wildlife Collection: page 22
Martin Harvey/The Wildlife Collection: Cover; pages 16-17
HPH Photography/The Wildlife Collection: Endpages; pages 11, 16-17, 20,
Robert Lankinen/The Wildlife Collection: pages 12-13
Tom DiMauro/The Wildlife Collection: page 18

Published by Rourke Publishing LLC

Copyright © 2002 Kidsbooks, Inc.

All rights reserved.
No part of this book may be reproduced or utilized in any form or by any means, electronic or mechanical including photocopying, recording, or by any information storage and retrieval system without permission in writing from the publisher.

Printed in the USA

Grambo, Rebecca
   Families / Rebecca L. Grambo
     p. cm – (Amazing Animals)
   ISBN 1-58952-146-3

FAMILIES

Written By
Rebecca L. Grambo

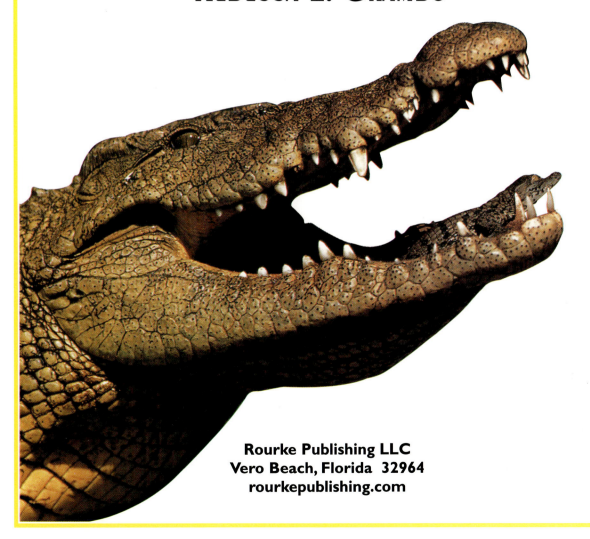

Rourke Publishing LLC
Vero Beach, Florida 32964
rourkepublishing.com

DIFFERENT FAMILIES

You may live with one parent or two. You may have a really big extended family—with grandparents, aunts, uncles, and cousins. Just like human families, animal families can be big or small.

In many bird families, both the mother and father take care of their chicks. They work hard to find enough food to feed their babies.

These sea turtles have just hatched from eggs their mother buried on the beach. Their mother will not raise them at all. Baby sea turtles are on their own, right from the start.

This zebra has just been born. It will depend on one parent, its mother, for food and protection.

In a wolf pack, usually only one pair of wolves has pups. The whole pack helps to raise these babies.

I'M HERE

Different animals have babies in different ways. Some animals lay eggs, which hatch baby animals. Others give birth to fully formed babies. Because humans are mammals, they give birth to their children.

Some snakes lay eggs like birds. Other snakes, such as this hog-nosed viper, give birth to live young.

Birds lay eggs. This chick has just hatched from its egg. It grew inside the egg for three and a half weeks. Some bird eggs are huge. An ostrich egg can weigh over three pounds.

Ants lay eggs, from which tiny wormlike larvae (LAR-vee) hatch. Adult ants take care of them until the larvae change into ants.

▲ Unlike most mammals, the echidna (ee-KID-nuh) lays an egg. This Australian animal has a pouch on its belly, like a kangaroo's. After laying the egg, the mother pushes the egg into the pouch. The egg hatches in about ten days, and, in another seven weeks, the very small baby emerges from the pouch.

BUNCHES OF BABIES

Sometimes a huge number of animals will gather in one place and have their babies at the same time. Many kinds of seabirds, such as these albatrosses, form nesting colonies called rookeries (ROOK-er-eez). There may be 10,000 pairs of albatrosses in one rookery.

Lions live in groups called prides. A lioness will nurse any of the cubs in her pride. At any given time, she may be feeding cubs from four different litters.

Baboons live in groups. The adults all share the job of caring for the young. Baboons will even defend babies that aren't related to them.

A bat mother has to find her own baby before she can feed it. She knows which is hers by its scent and voice. She may check 1,800 babies before finally finding the right one.

MOM ON HER OWN

For many animals, the only parent that offspring will ever know is their mother. A mother leopard raises her cubs by herself. The cubs stay with her, learning to hunt, until they are about one and a half years old.

Little hedgehogs stay with their mother until they are about seven weeks old. When they are first born, baby hedgehogs have some spines under their skin. Only after the babies dry off do the spines become prickly. More spines grow in later.

Polar bears give birth in the middle of winter in dens dug into the snow. Each cub weighs only about a pound and a half when born and will stay in the den for about three months. Cubs remain with their mother for more than two years, learning how to survive.

A mother scorpion carries her babies on her back for several days after they have hatched. If any fall off, she stops to look for them and then waits for them to climb back on.

DAD DOES IT

Some animal fathers play a big part in raising their young. This male midwife toad is carrying his future babies wrapped around his hind legs. He keeps the eggs damp so they don't dry out. When the tadpoles hatch, he will drop them into water.

A giant water bug, the female toebiter glues her eggs onto the male's back. He carries them around until they hatch into tiny toebiters.

This male jacana (juh-KAH-nuh) looks like he has feet growing out of his wings. Actually, he has a baby under there, tucked away for safety. A father jacana builds a nest, sits on the eggs, and then raises the chicks on his own.

The male Siamese fighting fish builds a nest by blowing bubbles. He guards the eggs and the baby fish that hatch from them. If the babies leave the bubble nest, their father sucks them up and spits them back in to safety.

Female seahorses lay their eggs into a pouch on the male's stomach. The father carries the eggs around until, one by one, the tiny seahorses pop out.

SHARING THE WORK

Animals that live in groups often share the work of taking care of the babies. Little baby elephants stay close to their mother or other female relatives. If the baby gets scared and cries, all the elephants in the herd come to the rescue.

Meerkats (MEER-cats) live in Africa. When it's time to eat, most of the meerkat members go off looking for food. But one or two adults stay behind to baby-sit the little meerkats.

Ants take very good care of their young. Sometimes the workers in an ant colony sit in the sun and get really hot. Then they go down to the nursery and warm the larvae and eggs with the heat stored in their body. Other ants work to keep moving eggs and larvae to the warmest places in the nest.

A baby mountain gorilla may sleep snuggled up with his mom or an aunt. He might play next to his great big dad, or share some food with his half-sister. When a new gorilla baby is born, all of the other gorillas are very interested in it.

READY OR NOT

Some baby animals need their parents to take care of them for a longer time than other babies do. A baby kangaroo, called a joey, is hairless, helpless, and less than an inch long when born. It immediately crawls up its mother's tummy and into her pouch, where it nurses and grows for six to eight months before ever coming out.

Only one day old, these baby mice can't hear or see, and they don't have any hair. But in only three weeks, they will be able to fend for themselves.

Opossums have about ten babies at a time. Like kangaroos, baby opossums spend a long time in the pouch. Even when they come out, after four to five weeks, the opossums stay with their mother for another eight to nine weeks.

▶ Baby flickers rely on their parents for everything. The chicks have to be fed for about four weeks before they are ready to leave the nest.

◀ A caribou (CARE-uh-boo) calf looks pretty wobbly just after it's born. But in only a few days it can keep up with the herd.

FEED ME

Baby animals need food to grow, and their parents give it to them in many ways. Mammal mothers, like this sow with her piglets, make milk in their body to feed their babies.

Dung beetle parents dig a shallow hole, then lay a single egg inside it. With the egg, they bury a ball of manure they have collected from animals who eat grass. When a larva hatches from the egg, it uses the manure as food until ready to change into an adult beetle.

Pelican parents eat fish and keep them in their stomach for a little while. When the baby pelican is hungry, mom or dad throws up a little bit of warm "fish stew." This food is easier for the baby to eat.

CARRYING THE KIDS

Animals may move their babies around to keep them safe. This mountain lion kitten wandered away. Now mom is taking it back to its brothers and sisters. Many mammals carry their young by the loose skin at the back of the neck.

Before her babies were born, this mother wolf spider carried a bag of eggs around with her. Now she is playing piggyback with her spiderlings. They will ride around on her back for about a week before they are on their own.

Life is a breeze, swinging in the trees—if you're a baby orangutan (uh- RANG-uh-tang). Holding on tight to its mother's long, orange fur, this baby really gets around. Orangutans rarely leave their home in the trees, living as high up as 100 feet above ground.

While its mother digs for lunch, this young giant anteater uses its claws to grip her fur. Many anteaters carry their young this way.

Snuggled on mom's tummy, this baby sea otter can nurse or nap. When the mother needs to look for food, she lets the baby float by itself.

TAKING SPECIAL CARE

Some parents do amazing things for their babies. Known as a pygmy marsupial frog, this mother is carrying her eggs under the skin of her back. The eggs will stay there until they hatch. Then the little frogs will come out through the skin.

Crocodiles make good mothers. When she hears noises from the eggs buried in her nest, the female crocodile tears open the nest. Then mom may take the eggs into her mouth to gently crack them. This helps the babies get out. Once the little crocs hatch, mom may give them a ride to the water in her mouth.

▼ During winter in the Antarctic, male emperor penguins spend two months with an egg on their feet, while mothers go off to feed. The fathers huddle in large groups on the ice, each keeping his egg warm. Going without food, a father penguin may lose 25 pounds before his mate returns.

Once a female hornbill lays her eggs in a hole in a tree, she walls herself up inside with them. She leaves a small opening so her mate can bring her food. When the baby hornbills get so big that there's not enough room for her, she breaks out and helps the father find food.

BIG BABIES

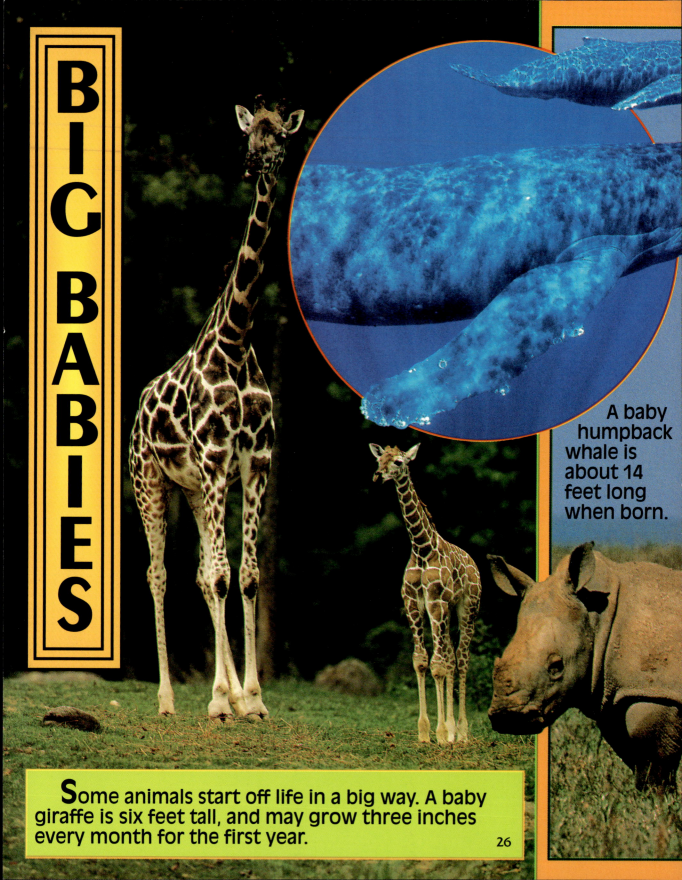

A baby humpback whale is about 14 feet long when born.

Some animals start off life in a big way. A baby giraffe is six feet tall, and may grow three inches every month for the first year.

Not all big animals start out big. A grownup mola (MOH-lah), or ocean sunfish, can weigh 4,000 pounds and be 11 feet long. But a baby mola is only one-tenth of an inch long when it hatches from its egg.

This baby rhino probably weighed about 143 pounds when it was born. That may seem big for a baby. But, in fact, this baby has a lot of growing to do. Its mother weighs about 3,500 pounds!

CLASS IS IN

Like us, baby animals learn a lot from their parents.

This baby manatee will stay with its mother for almost two years. It has to learn where to feed and rest, and how to find its way from place to place.

Black-tailed prairie dogs live in large groups. It's important for them to communicate with one another. These young prairie dogs must learn the difference between greeting calls and alarm calls.

This cheetah is showing her cub how to capture and kill prey. The cub must know how to hunt for itself in order to survive.

A young impala learns what is dangerous by watching how its mother reacts to things happening around her.

Raccoons eat lots of different foods. These little raccoons follow their mother, watching what she eats and how she finds it.